WEIRD ANIMALS

PLATYPUS

AMY CULLIFORD

A Crabtree Roots Book

Crabtree Publishing
crabtreebooks.com

School-to-Home Support for Caregivers and Teachers

This book helps children grow by letting them practice reading. Here are a few guiding questions to help the reader with building his or her comprehension skills. Possible answers appear here in red.

Before Reading:

- What do I think this book is about?
    - *I think this book is about a weird animal called a platypus.*
    - *I think this book is about where platypuses live.*

- What do I want to learn about this topic?
    - *I want to learn if a platypus would make a good pet.*
    - *I want to learn about what a platypus likes to eat.*

During Reading:

- I wonder why...
    - *I wonder why platypuses can live on land and in water.*
    - *I wonder why platypuses have such strange mouths.*

- What have I learned so far?
    - *I have learned that all platypuses have brown fur.*
    - *I have learned that platypuses use their big bills to eat.*

After Reading:

- What details did I learn about this topic?
    - *I have learned that all platypuses can swim.*
    - *I have learned that some platypuses can lay eggs.*

- Read the book again and look for the vocabulary words.
    - *I see the word* ***fur*** *on pago 6 and the word* ***bills*** *on page 8. The other vocabulary words are found on page 14.*

Look! I see
a **platypus**!

Platypuses can live on land and in water.

All platypuses have brown **fur**.

Platypuses use their big **bills** to eat.

Some platypuses can lay **eggs**.

All platypuses can **swim**!

Word List

Sight Words

a	eat	live	to
all	have	look	use
and	I	on	water
big	in	see	
brown	land	some	
can	lay	their	

Words to Know

bills

eggs

fur

platypus

swim

34 Words

Look! I see a **platypus**!

Platypuses can live on land and in water.

All platypuses have brown **fur**.

Platypuses use their big **bills** to eat.

Some platypuses can lay **eggs**.

All platypuses can **swim**!

Written by: Amy Culliford
Designed by: Rhea Wallace
Series Development: James Earley
Proofreader: Melissa Boyce
Educational Consultant: Marie Lemke M.Ed.

Photographs:
Shutterstock: vejrikluks: cover, p. 1; Lukas_Vejrik: p. 3; Mari_May: p. 5; Jaqui Martin: p. 7; John Carnemolia: p. 9; Chattanongzen: p. 11; slowmotiongli: p. 13

Crabtree Publishing

crabtreebooks.com 800-387-7650

In Canada: We acknowledge the financial support of the Government of Canada through the Canada Book Fund for our publishing activities.

Printed in the U.S.A./072023/CG20230214

Published in Canada
Crabtree Publishing
616 Welland Ave.
St. Catharines, Ontario
L2M 5V6

Published in the United States
Crabtree Publishing
347 Fifth Ave
Suite 1402-145
New York, NY 10016

Library and Archives Canada Cataloguing in Publication
Available at Library and Archives Canada

Library of Congress Cataloging-in-Publication Data
Available at the Library of Congress

Hardcover: 978-1-0398-0979-6
Paperback: 978-1-0398-1032-7
Ebook (pdf): 978-1-0398-1138-6
Epub: 978-1-0398-1085-3